THE BIG IDEA:
EINSTEIN & RELATIVITY

Paul Strathern was born in London. He has lec-
tured in philosophy and mathematics at Kingston
University and is the author of the highly suc-
cessful series *The Philosophers in 90 minutes*. He
has written five novels (*A Season in Abyssinia* won
a Somerset Maugham Award) and has also been
a travel writer. Paul Strathern previously worked
as a freelance journalist, writing for the *Observer*,
Daily Telegraph and *Irish Times*. He has one
daughter and lives in London.

Einstein
& Relativity

PAUL STRATHERN

ARROW

Published in the United Kingdom in 1997 by
Arrow Books

3 5 7 9 10 8 6 4 2

Copyright © Paul Strathern, 1997

First published in the United Kingdom
in 1997 by Arrow Books

Arrow Books Limited
Random House UK Ltd
20 Vauxhall Bridge Road, London SW1V 2SA

Random House Australia (Pty) Limited
20 Alfred Street, Milsons Point, Sydney,
New South Wales 2061, Australia

Random House New Zealand Limited
18 Poland Road, Glenfield
Auckland 10, New Zealand

Random House South Africa (Pty) Limited
Endulini, 5a Jubilee Road, Parktown 2193, South Africa

Random House UK Limited Reg. No. 954009

A CIP catalogue record for this book
is available from the British Library

Papers used by Random House UK Limited are natural,
recyclable products made from wood grown in sustainable
forests. The manufacturing processes conform to the en-
vironmental regulations of the country of origin

Typeset in Bembo by SX Composing DTP, Rayleigh, Essex
Printed and bound in the United Kingdom by
Cox & Wyman Ltd, Reading, Berkshire.

ISBN 0 09 923732 6

CONTENTS

INTRODUCTION

Einstein changed the universe, but died a failure. His Theory of Relativity established him as the greatest scientific mind since Newton. Relativity destroyed our notions of space and time, and brought into being a world which had previously been inconceivable. His celebrated formula $e = mc^2$ showed that matter could be converted into energy, heralding the nuclear age. And he made a major contribution to Quantum Theory. But in the end Einstein was unable to accept the implications of his discoveries – especially in Quantum Theory. As a result he wasted over a quarter of a century searching for a comprehensive theory which his own work had rendered impossible.

During the last half of his life Einstein became a public institution: 'the greatest genius in the world'. He accepted this absurdity with good grace, and put it to exemplary use – campaigning tirelessly against evils ranging from anti-semitism to nuclear weapons. The picture he presented to the world was that of the cliché absent-minded genius. The man himself was ambitious, very aware of his exceptional gifts, and ultimately a tragic figure. His social worth counted as little to him beside his failure to explain the ultimate workings of the universe with his Unified Field Theory.

LIFE AND WORKS

Albert Einstein was born of German-Jewish parents on the 14th of March 1879, in the small south German city of Ulm. His mother was the cultured daughter of a corn merchant from Stuttgart and enjoyed playing the violin. She was just 21 when Albert was born. His father Hermann was an affable, gregarious man with a large moustache who liked a glass of good German beer and enjoyed quoting poetry.

This was Germany under the 'blood and iron' rule of chancellor Bismark, when even the cab drivers wore uniforms. The Jews had been emancipated only in 1867, and in the year of Albert's birth the word 'anti-semitism' first appeared in a German magazine article.

A year after Albert was born his father's electrical goods business failed, and the family moved to the suburbs of Munich to live in the house of Hermann's brother Jakob. Here Hermann and Jakob established a small electrochemical works.

Albert was a distinctly slow, rather dreamy child. He had suffered a family disruption ('a fall from grace', as the psychologists call it), and he had a failed father. These are traits which occur with surprising frequency in the background of genius (Herr Johann van Beethoven the drunken singer, Mr John Shakespeare the unreliable glover etc), but otherwise Albert's early childhood was unexceptional.

Albert's father was not religious, and regarded himself as largely assimilated. As a result he sent young Albert to a catholic school, where he found himself the only Jew in his class. Much like everything else in the country, German schools were run on military lines. The teachers of the youngest children prided themselves on behaving like bossy, pedantic sergeant-majors. Young Albert was bored, learned little, and developed a deep grudge against authority which was to remain with him throughout his life. At home his mother made him study the violin, which he enjoyed and learned to play well – another trait which was to remain with him for life. Albert's father was mainly pre-occupied in trying to keep the family business afloat through an economic recession, but he did make sporadic attempts to interest his son in vaguely academic matters. One day he showed his son a compass. Albert asked why the needle always pointed in the same direction. Hermann explained that this was due to magnetism. But how did the magnetism manage to cross space, Albert wanted to know. For this Hermann had no answer.

That night Albert lay awake pondering how an invisible force could pass through space.

At the same time, 'Onkel Jakob' introduced the young boy to algebra. 'It is a merry science,' he explained. 'When the animal we are hunting cannot be caught, we call it x temporarily and continue to hunt it until it is bagged.' Bertl ('little Bertie', his family nickname) was soon hooked.

In 1891, when Einstein was 12, another amateur teacher arrived on the scene. In those days it was the custom amongst Jewish families in central Europe to invite an impecunious member of the community to dinner on Thursdays. The Einstein household entertained Max Talmey, a medical student. Max began lending young Bertl books on popular science, which his otherwise largely idle brain quickly devoured. Here again, Einstein developed a trait that lasted for the rest of his life. He was largely self-taught, paying little or no attention to his teachers. He preferred to follow his own interests, and do things his own way. The result was an exceptional depth of knowledge, accompanied by frequent difficulty with even the most elementary exams.

Max Talmey was soon bringing Einstein books on plane geometry, and in no time the boy was teaching himself calculus. Each week Max would check on young Albert's progress, until eventually he was forced to admit: 'I could no longer follow him.' Vainly Max encouraged him to read books on medicine and biology, but Albert wasn't interested. They provided insufficient intellectual challenge: he appeared only interested in trying to comprehend complex notions, and seeking the underlying principles involved.

So the ageing medical student now introduced Albert to his own favourite subject: philosophy. The young teenager suffering from 'learning difficulties' at school began studying the works of

Kant. These are fiendishly difficult: German metaphysics at its most prolix and obscure. Indeed, there may even have been an element of maliciousness in Max's move, intended to put Albert in his place. But Kant's work contained the greatest philosophical system of them all, a construction of exceptional profundity which sought to explain absolutely everything. Before, Einstein had encountered intellectual subtlety and refinement, concepts which required extreme concentration even to grasp, and techniques of brilliance. But here he learned, for the first time, what the mind in all its glory was capable of achieving: a system which embraced the universe. Einstein never forgot this lesson. Max's joke, if it was one, was to rebound with a vengeance.

In 1894, when Einstein was fifteen, his father's business failed once more. The family moved to Italy, where his father set up a new factory near Milan. But Albert was left at a boarding house in Munich, so that he could finish his diploma at the Luitpold Gymnasium. This would enable him to enter university, where he would take a degree in engineering, and then join the family business. His studies were to be supported by his mother's family, until such time as Hermann got back on his feet.

Within six months Einstein had a nervous breakdown and was expelled from the gymnasium because (according to his report) his presence in class was 'disruptive and disturbs the other pupils'. The mental breakdown may have been faked, so that he would be sent to join his parents in Italy. But the expulsion seems to have been real enough. Einstein had developed an aversion to discipline, and according to his memoirs regarded the school curriculum as a concoction of deceit, irrelevancy and boredom. In ancient Greek, history, geography, and surprisingly biology and chemistry, he didn't even bother to try. He was becoming aware of his precocious intellect (which outstripped all in maths and physics), and this gave him a pronounced self-assurance. This, combined with a degree of immaturity, made him appear cocky and insolent.

Albert now spent a very enjoyable year in Italy. He didn't go to school, but spent part of his time writing a paper about one of the most difficult scientific problems of the day – the relationship between electricity, magnetism and the ether (the invisible medium which transmitted electro-magnetic waves). On a professional level this had nothing original to say, but it was a remarkable feat for a sixteen-year-old. It also showed that he was still thinking about magnetism and how it travelled through space.

At the end of the year he took the entrance exam to the Eidgenössische Technische Hoch-schule in Zurich (generally known as the Zurich Polytechnic). The professor of physics, Heinrich Weber, was astonished at his remarkable marks in maths and physics. His father Hermann reacted somewhat differently when he learned of his marks in French, biology, history and various other subjects. Einstein had failed resoundingly – and almost certainly on purpose. He had no wish to embark upon an engineering course which would end up with him entering his father's electrical business. However, as a result of Pro-fessor Weber's personal intervention Einstein was offered a place at Zurich Polytechnic for the following year. There was just one condition attached: Einstein would have to attend a school, any school, during the intervening year.

Hermann recognised his son's reluctance to

enter the family business. But he was at a loss. He had no money. Should he insist on Albert coming to help in the business right away? Once again he contacted his wife's relatives, and once again they agreed to finance Albert's education. But this time they wanted to see some results. There was no sense in throwing away good money on a mere wastrel.

Long after all the disappointments and disagreements, many years after Hermann's death, Einstein always insisted upon one thing about his father: he was 'wise'. Hermann's understanding of his wayward son's needs and intuitive aversion to burying himself in the family business epitomised Hermann's wisdom. Without this, there would have been no Relativity.

Hermann decided to send Albert to school in a village outside Zurich, and agreed to let him study physics and mathematics, rather than engineering, when he entered Zurich Polytechnic. And this time, instead of staying in a lonely boarding house, Einstein would stay with the family of one of the teachers.

Despite his father's 'wisdom', Einstein still had misgivings about going back to school. But these were soon allayed. Aurau turned out to be a pleasant riverside spot amidst rolling vineyards, and his hosts the Winteler family were lively and welcoming. This was not Germany, it was Switzerland. Instead of pedagogic rigidity, he found openness to intellectual debate. Einstein thrived, joining the family at weekends on bird-watching expeditions and hikes into the mountains.

Einstein had first learned the violin from his mother, and by now was an accomplished amateur player. During musical evenings at the Winteler home he would entertain the family by playing duets with their 18-year-old daughter Marie, who would accompany him on the piano. Albert was a spirited violinist: music seemed to bring out the passionate side of his nature.

Photos from the period show Einstein as a handsome young man with dark curly hair, a budding moustache and a confident air. He is well-dressed, despite incipient evidence of the dishevelled appearance which was later to become his hall-mark. At this early stage, it merely adds a touch of debonair raffishness. Perhaps inevitably, Marie fell in love with him.

This was Albert's first romantic experience. It

seems to have been fairly intense, platonic, and a little one-sided. Mathematical-physics was already Einstein's overriding passionate interest, as it was to remain so throughout his life. But he liked the company of women, and he knew he was attractive to them.

During this period Einstein had a lively and self-confident social manner, with a definite edge to it; and he enjoyed laughing uproariously. A schoolfriend remembered how he adopted the demeanour of 'a laughing philosopher and his witty mockery lashed any conceit or pose'. But he had this incorrigible tendency to drift off into his own thoughts. Perhaps inevitably, the romance with Marie foundered when he took up his place at Zurich Polytechnic in the autumn of 1895. She was heartbroken; he caddishly decided to forget all about it.

The Zurich Polytechnic was the finest technical school in central Europe at the time. Its laboratories were superbly equipped by Siemens (ironically one of the big conglomerates which was responsible for putting Hermann out of business); and it attracted teaching staff of the highest calibre. Despite this, Einstein seldom turned up for lectures. One of his professors, the great Russian-German mathematician Hermann Minkowsi, referred to him as 'a lazy dog'. But Einstein remained as cocksure as ever. His ingratitude towards Professor Weber – who had been responsible for getting him into the college – was characteristic. Weber's physics lectures included few of the great technical advances made during the previous 20 years, and Einstein was frankly dismissive. In the lab he refused to obey instructions, preferring to devise his own more up-to-date methods. During an experiment to determine the effects of the ether his equipment blew up, giving his right hand a nasty injury. Fortunately this was not lasting, and he was soon able to take up the violin again.

For the most part Einstein spent his time reading avidly – working his way through the latest advances in physics. During the previous century enormous progress had been made in science – especially in physics, which now appeared to be the cutting edge of scientific knowledge. Things had reached the stage where all the disparate elements of scientific knowledge seemed to be coalescing into a vast comprehensive overview. The prospect of humanity achieving a total knowledge of the world lay on the horizon. Many felt that this was the most exciting time in history for a scientist to be alive. Future generations would have nothing to discover, and be condemned to the mere drudgery of measurement.

Yet in other quarters doubts were beginning to appear. A number of the old certainties were coming into question, leading to a growing suspicion that classical physics was not adequate to describe the increasingly complex realities of the physical world.

Such thoughts were uncannily echoed by many intellectuals in Zurich who viewed the world from a non-scientific point of view. Trotsky, Lenin, Rosa Luxemburg (and later the Futurists, Dadaists and James Joyce) all frequented the cafés of Zurich. In those days Zurich was much more than a provincial banking centre run by gnomes. It was a lively city at the heart of Europe, with a cosmopolitan café society.

In between bouts of increasingly esoteric reading and research, Einstein would meet his university friends at the Café Metropole, a popular student haunt down near the river. He had started smoking a pipe, and his favourite drink was iced coffee. (In those days Einstein didn't drink beer, mainly because he didn't have enough money. Indeed, throughout his life, he was never keen on alcohol, as he felt that it dulled his brain.)

Einstein had a small circle of close friends. All of them were bright, studying maths or physics, and obsessed with the ultimate questions of science. Without such qualities, it would have been impossible to follow their conversations.

Marcel Grossman was perhaps the first among Einstein's classmates to recognise that his brilliance was something out of the ordinary. When it came to exams Grossman would selflessly lend Einstein his lecture notes – the only way he could have covered the syllabus in time. Einstein's fellow engineering student Michelangelo Besso was a likeable character, who shared Einstein's interest in philosophy. It was he who introduced Einstein to the works of Ernst Mach, the philosopher of science whose name is now immortalised in the measurement of the sound barrier. In those days, Mach was seeking to have a similarly cataclysmic effect on the unquestioned abstract assumptions of classical physics. A third close friend was Fritz

Adler, son of the founder of the Austrian Social Democratic Party. Einstein admired Adler for his unflinching idealism. Einstein's iconoclasm was more than just youthful kicking over the traces. The anti-authoritarianism, the anti-militarism, the disdain for outmoded methods and assumptions – all these were part of a growing social idealism. This was profoundly held, if somewhat naively utopian at times: two characteristics his idealism was to retain throughout his life.

When not working on his own in his room, or in earnest conversation (punctuated by bursts of manic laughter) with his friends at the Café Metropole, Einstein went sailing with his landlady's daughter on Lake Zurich. This marked the beginning of two hobbies which he pursued with enjoyment to the end of his days – namely, sailing and flirting. (The landlady's daughter was not the only one invited on intimate sailing trips, even though the boat did belong to her family.)

Only one person was capable of encompassing all these facets of Einstein's life. This was Mileva Maric, the sole woman student in his class. Mileva was a Serbian from Novi Sad, then part of the Austro-Hungarian Empire. Her father was a civil servant, who had sent her to the Zurich Polytechnic because as a woman she wasn't allowed to study advanced physics at home. (Not until *eight years later* did Marie Curie become the first woman in France to receive a doctorate *in any subject* – the same year as she won her first Nobel Prize.) Unlike the other women with whom Einstein spent his time, Mileva was rather plain and not in the least flirtatious. She seldom laughed, and owing to a permanently dislocated hip walked with a slight limp. Many have wondered how she came to occupy such a central role in Einstein's life at this time. Contemporary photos show her as having heavy slavonic features, but the cast of her dark eyes and wide lips hints at a covert sensuality. She was also the first woman Einstein had encountered with whom he could discuss his deepest concerns. When he began speaking about physics, she knew enough of what he was talking about to make suggestions. And Einstein certainly admired her pioneering independence, a rare achievement amongst women of the period.

In 1900 Einstein borrowed Grossman's lecture notes for the last time and took his final exams.

His results were uneven, barely hinting at his exceptional scientific brain. These results, combined with his refusal to listen to his teachers, ensured that he received no backing for the academic career he wished to pursue. He applied to several universities – in his own fashion. Einstein's characteristic blend of unorthodoxy and intellectual self-esteem (with little concrete evidence to support this) ensured that he received no offers.

In 1900 Einstein became a Swiss citizen, partly because he felt at home there, but also to avoid having to return to Germany for military service. But this didn't help with his job applications, which also suffered from another personal characteristic that was even more undisguisable than his self-esteem. Namely, his Jewishness. Antisemitism was rife in the professions throughout Europe. (It was only six years since the notorious Dreyfus Case had rocked Paris – when Dreyfus, a Jewish officer in the French army, was sentenced to Devil's island on a trumped-up spying charge.) Einstein's money began to run out, and he finally had to accept a temporary post at the technical school in Winterthur, just ten miles north of Zurich.

Einstein's attitude towards discipline ensured that he was a popular, though ineffectual teacher. During his free time he continued with his researches, which had now begun to focus on the possibility of a link between molecular forces and the force of gravity which acted over vast distances. At this stage it appears he was trying to incorporate the latest scientific advances into the overall structure of classical physics, rather than suggesting any alternative structure. But already he was beginning to ponder on the larger scheme of things – trying to improve on Newton. Ambitious stuff, but as he commented in a letter to Grossman:'It is a wonderful feeling to recognise the unifying features of a complex of phenomena which present themselves as quite unconnected to the direct experience of the senses'. He was beginning to discover his strengths.

When he could, he would travel to Zurich at the weekends to see Mileva, and during the week they exchanged letters. After one weekend in May, Einstein wrote: 'How delightful it was last time when I was allowed to press your dear little person to me in the way nature created it'. They had become lovers.

After a few months Einstein's teaching job came to an end, with no prospect of further work. Hearing of this, his old student friend Grossman asked his father to recommend Einstein for a job at the Swiss Patent Office in Berne. Einstein learned that there were no jobs available at present, but he would be considered if a vacancy arose. The next thing he received was the news that Mileva was pregnant.

Einstein was now 21, out of a job, and had practically no money. His father's business had once again gone bust, and his mother's relatives were no longer willing to support him. Einstein told Mileva they would get married, but they both knew this was impossible. He was incapable of supporting her.

Eventually Mileva travelled back to Novi Sad, where she gave birth to a girl – whom Albert and Mileva referred to in their letters as 'Lieserl' ('little Lisa').

Early in 1902 Einstein travelled to Berne, where eventually there was a vacancy at the Patent Office. He became a technical examiner (Third Class), which involved sorting through the various inventions submitted for approval by the Office.

These included the usual range of ingenious gadgets, hilarious implausibilities, and simple devices upon which financial dynasties would be founded. Einstein would examine each gadget, and then read through the accompanying submission (often as tortuously complex and impenetrable as the device it purported to describe). His task was to ensure that these two disparate elements bore some relation to one another, and that at least one of them was comprehensible. He discovered that even the most complex notions could usually be reduced to a set of simple fundamental principles. This was a lesson he would never forget.

Meanwhile Mileva and baby Lieserl remained 600 miles away in Novi Sad. It appears that Lieserl was a sickly child, and Mileva was none too well herself. The ensuing story of Lieserl and her parents is a typical tragedy of the period, which only came to light in the 1990s and still remains to be fully recounted. It appears that Mileva's parents prevailed upon her to have Lieserl adopted, and there was little Einstein could (or would) do about it. So what became of the first-born child carrying the genes of one of the greatest scientific minds of all time? Lieserl appears to have vanished without trace – apart from one curious episode. Over 30 years later, when Einstein was world famous and living in America, he heard that a woman was trying to pass herself off in European society as his illegitimate daughter. Einstein refused to issue a disclaimer, and instead quietly hired a private detective to discover if the woman's claim was genuine . . . The end of this story is not yet fully known. In the normal course of events, Lieserl could have been expected to live at least into the 1970s. The editor of Einstein's papers, Dr Robert Schulmann, has hinted that new evidence may come to light after the troubles in the former Jugoslavia have been resolved.

In December 1902, less than a year after the birth of her daughter, Mileva Maric left Novi Sad alone and travelled to Switzerland. It was obvious to all her friends that she had suffered some profound sadness, but she never revealed its cause. Out of a mixture of compassion, affection and duty Einstein had decided to marry Mileva. Her motives were equally mixed, but she felt that she had nowhere else to turn.

On 3rd January 1903, Albert and Mileva were

married. After a celebratory meal with a few friends at a local restaurant, the newly-weds set off into the freezing night for Einstein's small flat at 49 Kramgasse, just around the corner. When they arrived, Einstein found he had somehow mislaid his key. This is usually recounted as a classic example of Einstein's absent-minded eccentricity. Others may be inclined to a more Freudian interpretation.

Einstein was now 23, and desperately poor. In order to avoid facing up to this difficult reality, he took to burying himself in his scientific researches. This was to be a recurrent feature: when the going got tough, Einstein would escape into his abstract world. During this period he produced a number of scientific papers, some of which were printed in the prestigious *Annalen der Physik*. Einstein was concerned with thermodynamics, and he also developed certain statistical methods for assessing the movements of the vast number of molecules which occupy a comparatively small volume of liquid or gas. None of these papers are particularly original, and only with hindsight is it possible to see hints of the great discoveries to come.

In 1904 Mileva gave birth to their first son, Hans Albert. Some months later Einstein's old Zurich pal Besso also got a job at the Patent Office. This meant that he now had someone with whom he could discuss his scientific researches. His ideas were now extending beyond Mileva's range, and their discussions were even more severely curtailed by her motherhood. Not unnaturally, Mileva resented this, and Besso was not always welcome at 49 Kramgasse. Instead, Einstein took to discussing his ideas with Besso as they walked back from work, often by a circuitous route.

Einstein's published papers may not have been of the greatest importance, but the range of his preoccupations and insights was decidedly original. So original in fact, that he could see no way of expressing them in any coherent form, as he complained to Besso on their walks. By now Einstein had realized that classical physics was finished. Space, time and light did not conform to Newton's definitions. An entirely new explanation of the universe was needed.

Such were the revolutionary ideas forming in Einstein's head, and he spent as much time as he could trying to elaborate them. Yet life in the Einstein home was hardly conducive to thought as penetrating and comprehensive as any since Newton. Nowadays the flat at 49 Kramgasse has the aseptic calm of a small museum devoted to Einstein and the theory of Relativity which he conceived here. During those early days, in the heat of creativity and family life, the atmosphere was somewhat more heady. Visitors recalled the heavy stench of drying clothes and nappies, Einstein's pipe tobacco, and smoke seeping from the leaking stove. In winter it was too cold to open the windows; in summer the heat intensified the smells. Einstein would be found immersed in a book, absent-mindedly rocking the howling baby in a cradle with his foot, while Mileva washed up at the sink. Occasionally his friends would come across him leaning over amongst the pavement throng, his notebook spread open on the baby's pram, immersed in some long calculation while the baby hit him on the head with his rattle.

All this obsessive thought came to a sudden and spectacular climax in 1905. This was to be Einstein's *annus mirabilis*. During the course of this year he sent four papers to the *Annalen der Physik*. These literally changed the world.

The first paper to appear in *Annalen der Physik* was 'On a Heuristic Viewpoint Concerning the Production and Transformation of Light' (*Uber einen die Erzeugung und Verwandlung des Lichtes betreffenden heuristischen Gesichtspunkt*). Einstein himself considered this 17-page paper 'very revolutionary', and it was indeed to transform our entire understanding of the nature of light – in such a way that physics would never be the same again.

In order to understand the importance of Einstein's paper, we must first trace the scientific history of light. Since the time of the Ancient Greeks, philosophers and scientists had believed that light consisted of tiny grains of matter. With the invention of the telescope in the early 17th century, this view came under question. In 1678 the Dutch astronomer and physicist Christian Huygens suggested that light was in fact composed of waves. But as one contemporary critic objected: 'How even can the waves of the sea travel without brine?' In other words, waves always require a substance or 'medium' to transmit them. Light waves could travel through the air, water, and glass, but how did they travel through space or a vacuum? Huygens came up with the idea of an all-pervasive invisible substance called the ether. Later this was elaborated as a weightless, static entity that pervaded the entire universe.

In 1704 Isaac Newton published his great work on light entitled *Optiks*. This described in exhaustive fashion all manner of behaviour and qualities of light. To account for these multifarious properties he proposed a corpuscular theory, with light made up of particles which were in some way influenced by waves. Unfortunately Newton was unable to come up with a convincing explanation which blended these two apparently contradictory elements.

The wave theory of light received a great boost in the following century from work conducted by the Scottish physicist James Clerk Maxwell, who was to die in 1878, the year before Einstein was born. In the 1860s Maxwell calculated that both electric and magnetic forces ought to move through space at around the speed of light. He immediately inferred that light too was a form of electromagnetic radiation, transmitted in waves through the ether. He also maintained that the wave-length of light occupied only a small range of the spectrum of electromagnetic waves, and predicted that other types of electromagnetic waves with different wavelengths would soon be discovered.

These findings were confirmed in 1888 by the German physicist Heinrich Hertz, who discovered radio waves. These behaved in precisely the same fashion as heat and light, all exhibiting wave-like qualities. Hertz was the first to broadcast and receive the newly-discovered radio waves, but unfortunately died of blood-poisoning in 1894 before he could find a use for his discovery. It was the Italian-Irish physicist Marconi who was to develop the practical application of Hertz's discovery. Hertz's work further corroborated the wave theory of light by confirming Maxwell's suspicion that electric and magnetic forces travelled through the ether at the same speed as light.

Unfortunately, some were now beginning to have doubts about the ether, upon which the entire wave theory of light appeared to rest. Not only did the ether have to fill all space and permeate all bodies, but it had to be consistently uniform and rigid if it was to transmit light waves.

In 1887 the American naval scientist Albert Michelson and his collaborator Edward Morley conducted an experiment intended to measure the velocity of the earth. This they hoped to do by showing the effect of the earth's motion through the static ether. But they discovered there was no effect, which led observers to begin questioning the existence of the ether. Despite its all-pervasiveness, had anyone ever found any real evidence for the existence of this elusive substance? Had anyone even registered its presence in an experiment? Yet without ether, what medium was there to convey light waves through space?

Despite all Maxwell's work, and Hertz's apparently conclusive confirmation, new evidence now began to appear which seems to contradict the wave theory of light. A photo-electric effect was observed when certain solids were struck by light. This effect was caused by an emission of electrons. In particular, it was found that when ultraviolet light struck certain metals it caused measurable electron emission. The German physicist Phillipp Lenard explained this by suggesting that these photo-electrons, as they were known, were displaced from the metal by the falling light. If this was the case an increase in light would surely bring about an increase in the speed of the electrons dispersed. But this didn't happen. Instead, a greater number of electrons was emitted, but at the same speed. Then Lenard discovered something even odder. When he altered the colour of the light (or, in other words, changed its wave-frequency) *this* affected the speed of the emitted electrons. As the frequency was increased, so was the speed of the emitted electrons.

These, together with related phenomena which indicated discrepancies in the wave theory, were investigated by the German physicist Max Planck in Berlin. He embarked on a mathematical description of these phenomena which led to increasingly startling results. These appeared to contradict the basic principles of classical physics as they had been understood since the time of Newton, two hundred years previously.

On 14 December 1900 Planck reached a momentous conclusion. Later that day, as he was walking in the Grunewald woods near Berlin, he announced to his young son: 'Today I have made a discovery as important as that of Newton . . . I have taken the first step beyond classical physics'. According to Planck's findings, when light struck matter it was not absorbed or emitted in a continuous flow, as suggested by common sense (and both the wave and particle theories of light). Instead it was emitted or absorbed in separate bursts of energy, much like particles. These separate bursts of energy he called quanta, from the Latin 'how much'. The size of these quanta was related to the wave-frequency of the light.

Light appeared to be self-contradictory, consisting at the same time of both waves and particles. This was literally inconceivable, and Planck refused to go along with it. He maintained that his theory only described the *relationship* between light and matter. It did not apply to the *nature* of light itself. He felt sure that the discontinuous bursts of energy – the quanta – somehow joined up to become waves, when travelling apart from matter. But he was unable to explain how this happened. Quantum Theory had begun, but even to its originator it remained largely inexplicable.

It was Einstein who eventually suggested a solution to this problem. Planck, much like his quanta, was two contradictory things at once: he was both right and wrong. Quanta did explain light in relation to matter; but more than that, it also explained the very nature of light itself. Einstein formulated his idea in precise physical-mathematical argument in his historic 1905 paper 'On a Heuristic Viewpoint Concerning the Production and Transformation of Light'. Einstein himself may have considered this 'very revolutionary', but he remained oddly tentative. He admitted that his view was 'irreconcilable with established principles . . . perhaps even ultimately untenable'.

According to Einstein, light should be treated for some purposes as independent particles much like a gas, but with zero mass at rest. Light in such cases consisted of quanta (later to be called photons). But in addition there were other cases when light would exhibit wave-like behaviour, and should then be treated as if it consisted purely of waves.

Planck had detected a new anomaly, extending beyond the range of classical physics. Einstein's solution meant the definitive end of the classical view of physics where light was concerned. Worse than that it defied the rules of logic. Light was expected to be two contradictory things at the same time. How could something consist of discreet particles, and yet at the same time be a wave, with a measurable wavelength? Science had entered a new era, beyond common sense. In cases such as this, Science didn't necessarily seek to understand what was going on, it sought to describe it – in ways (whether contradictory or not) which could be *used* to explain phenomena and determine future knowledge.

Einstein's 'heuristic viewpoint' (based on observed phenomena, rather than comprehensive theory) explained the photo-electric effect, and it explained away the ether. Light travelling as quanta behaved like particles, which unlike waves required no medium to transmit them. As many had suspected for some time, there was now no need for this elusive substance. (Einstein's badly-assembled university experiment to detect the ether may have blown up and damaged his hand, but now he had finally got his own back by abolishing the ether altogether.)

Einstein's new theory of light also explained certain anomalies which had arisen in classical physics. Such a mechanical view of the world was evidently not ultimate. Although Einstein's view of light bore a curious resemblance to Newton's imprecise formulation 200 years previously, it spelt the beginning of the end for Newton's physics. Einstein's physical–mathematical argument set the scene for Quantum Theory, and made Planck's original conception (quanta) fundamental to the very nature of light.

However, Planck didn't see it this way. He stuck to his guns, maintaining that quanta referred only to light in relation with matter. As late as 1912 Planck was still attacking Einstein's 'heuristic viewpoint' in his lectures at Berlin University. And he wasn't the only one. Few scientists were willing to believe that science could possibly defy logic in this fashion. Not until 1915 did Einstein's theory of light start to gain acceptance, as the experimental evidence in its favour became more and more undeniable. By the 1920s Quantum Theory was beginning to emerge as one of the major break-throughs of the twentieth century. Planck was awarded the Nobel Prize in 1919, and Einstein two years later. (Einstein received this ultimate accolade for his work on light and quanta, not relativity.) In the practical field, Einstein's theory of light was to play a leading role in the development of television. But its most noticeable application today is in the 'electric eye' which automatically opens doors. As a child Einstein had lain awake in bed wondering how magnetic force passed through space: 20 years later, his explanation of this phenomenon transformed physics.

'A New Determination of the Size of Molecules' (*Eine neue Bestimmung der Molekuldimensionen*) is Einstein's second paper to appear in the famous Volume 17 of the *Annalen der Physik* (a rare copy of which is rumoured to have changed hands recently for over $10,000). Einstein's second paper outlines a method for describing the size of a sugar molecule. This work has correctly been described as 'a minnow among the whales of the other three papers'.

Having completed this diversion, Einstein returned to more fundamental matters. His next paper was entitled 'On the Motion of Small Particles Suspended in a Stationary Liquid, according to the Molecular Kinetic Theory of Heat' (*Uber die von der molekularkinetischen Theorie der Warme geforderte Bewegung von in ruhenden Flüssigkeiten suspendierten Teilchen*). The study of murky liquid would hardly seem a promising area for making earth-shattering scientific discoveries, but Einstein's propensity for reaching to the very root of a problem made it otherwise.

Once again, a little history is needed. In 1828 the Scottish naturalist Robert Brown began observing pollen dust suspended in water, during the course of his botanical researches. When he studied this under a microscope, he noticed that the individual particles of pollen exhibited continuous zigzagging, apparently random movements. It was as if they were alive. But when he substituted an inorganic powder for the organic pollen, he observed precisely the same effect. Brown appeared to have stumbled across an example of that scientific impossibility: perpetual motion. Brown remained baffled by this phenomenon, named Brownian Motion after him, which continued to puzzle the scientific community throughout the 19th century.

When Einstein studied Brownian Motion he was intrigued by this apparent defiance of the laws of physics, and came up with a characteristically original and daring solution. According to the Molecular Kinetic Theory of Heat the invisible molecules of the liquid were in motion, this motion intensifying as the liquid rose in temperature. In Einstein's view, the seemingly random behaviour of the suspended particles was in fact due to them being bombarded by the invisible molecules which made up the liquid. This was a particularly daring suggestion, as many reputable scientists still remained to be convinced that molecules and atoms actually existed. (These entities still defied all attempts at observation. Just like the elusive ether, no one had yet seen a molecule.) But Einstein now went one stage further, and set about *proving* the existence of these unseen molecules. Making use of statistical dynamics, he even set about predicting the precise number of molecules in any given quantity of liquid.

The simplest outline of how he managed to do this gives an inkling of the complexities involved.

An object in water (or in any liquid or gas) suffers continual bombardment from the molecules of that liquid or gas. By the workings of chance, the amount of molecules bombarding a large object from all sides will even out – and the object will not be displaced. However, a much smaller object such as a pollen particle is liable to be pushed first in one direction then another, owing to the slight excess of molecules bombarding it from any one direction. Einstein came up with a formula to describe this effect. According to this, the average displacement of the visible particles in any single direction increased as the square root of the observation time. If the distance covered by the particles in this time was measured, it was then possible to calculate the number of invisible molecules within a certain volume of liquid or gas. In this way Einstein calculated that one gram of hydrogen contains 3.03 x 10^{23} (ie, over 3 million million million million) molecules.

Einstein's paper not only set out to prove the existence of molecules, but also to show the density of their occurrence and how to map their behaviour.

Einstein's theoretical demonstration was to be confirmed three years later in practical experiments conducted by the French physical chemist Jean Perrin. The experiments conducted by Perrin on the Brownian Motion of gamboge (a yellowish resin) in water were the first practical demonstration of the physical existence of atoms. His experiments also revealed the remarkable accuracy of Einstein's purely theoretical calculations.

This practical confirmation of Einstein's work highlights an essential feature of his methodology. Here was the new scientific approach of the 20th century, as characteristic in its way as Cubism and atonal music. The 19th century had witnessed the growth of many branches of science from infancy to full maturity. During this period scientific method had been largely empirical. Dramatic advances had been made through experiment, observation and the use of ingenious apparatus. But Einstein's method was not experimental. On the contrary. He remained at heart an unrepentant theorist. Experiment would come later, revealing facts which conformed with his theories. The old method of constructing theories out of facts supported by experimental evidence was far too slow and prosaic for Einstein. His mind preferred to race ahead and confront ultimate possibilities far beyond the range of experiment.

Einstein was not alone in taking such an approach. This was to be the method of the coming century. (Atomic explosions and rockets to the moon were proved in some detail to be theoretically possible long before their actuality.) The genius with the slide rule now stood at the cutting edge of science, rather than the boffin in the lab.

In his previous papers Einstein had demonstrated the nature of light and the existence of atoms, two fundamental entities. In doing so he had transformed the entire way in which science viewed the world. These unique insights would have been enough to establish him as one of the leading scientific minds of his age. But he now went one step further. He combined his insights into these microcosmic worlds and produced a macrocosmic theory which transformed the universe. This achievement was to establish him as one of the most creative intellects in human history (alongside the likes of Newton and

Beethoven).

Einstein was only 26 years old, and to all appearances just an impecunious minor functionary in the Patent Office at Berne. He may have produced the occasional scientific paper in his spare time, but he was by no means recognised even by the local academic community.

Throughout his *annus mirabilis* Einstein worked in virtual isolation. Just over two centuries previously, Newton had experienced a similar year of miraculous creativity, during which he also produced much of his major work. He was around the same age as Einstein, and was in flight from the plague, living in isolation in the country. But Newton didn't have to go to work every day, and didn't live in a small apartment with a wife and baby. Einstein's intellectual feat appears to be unparalleled in the history of the human mind. The justification for such hyperbole only becomes apparent with the advent of his fourth paper.

For some time now Einstein had been pondering the question of how to achieve certainty in physical formulations of great magnitude. Surely there had to be some ultimate unvarying yardstick against which all varying quantities could be measured. If not, everything simply became relative – depending upon the frame of reference from which it was viewed.

Contrary to the popular image, Einstein did not simply meditate on these matters in a profound contemplative trance (during which all manner of absent-minded eccentricities took place). Einstein preferred to cultivate this image amongst his colleagues, but the truth was not so cosy. He was a passionate thinker, who privately admitted that long periods of profound theoretical thought brought about in him a 'psychic tension . . . visited by all sorts of nervous conflicts'. By the spring of 1905, Einstein found himself plunged into the most turbulent mental crisis he had ever experienced.

Walking back from the Patent Office, he would test out his ideas on Besso, but it soon became apparent that his thoughts were drifting into a realm where Besso's comments were of no use to him. Einstein took to pacing the medieval streets and arcades of Berne, often absent-mindedly wandering across the river and far out into the surrounding countryside. On one occasion he only came to when he found himself walking down a country lane soaked to the skin in the midst of a thunderstorm.

Part of Einstein's exceptional talent lay in his ability to think through the most complex formulae and problems to the fundamental principles which underlay them. From these he would try all manner of deductions, in the search for further, even more fundamental principles. Through the spring of 1905 the constant mental effort involved in all of this brought Einstein to the brink of mental collapse. He was exhausted, both mentally and physically. He couldn't eat properly, and he couldn't sleep properly. And worse still, no matter how he focused his mind, his thoughts remained fragmentary. Try as he might, the pieces proved incompatible. They refused to coalesce into any consistent theory – which he felt sure was there. Somewhere. He had reached an impasse: there appeared to be no way forward. Walking back from the Patent Office with Besso one day, he finally confessed: 'I've decided to give it up – the whole theory.'

That night he went to bed in despair, at the same time aware of a curious sense of relief. He lay in a daze, neither awake nor asleep. Next

morning he came to in a state of extreme turmoil. 'A storm broke loose in my head,' was how
he described it. And in the midst of this storm he
suddenly grasped the idea that had eluded him
for so long. In his own words, it was as if he
gained access to 'God's thoughts'.

This was no personal communication with the
Divine. Einstein always insisted that he didn't
believe in a personal God. But in common with
many leading minds of his era (such as Picasso,
Wittgenstein and even Freud on occasion), he
continued to use the word God in association
with great truths which lay at the very limit of
human understanding. It alone appeared to evoke
the sense of awe involved. Einstein and Picasso, it
seems, both experienced that profound sense of
wonder spoken of by the great philosophers from
Plato to Kant.

Einstein described what he had understood as
follows: 'The solution came to me suddenly, with
the thought that our concepts and laws of space
and time can only claim validity insofar as they
stand in a clear relation to our experiences; and
that experience could very well lead to the alter-
ation of these concepts and laws. By a revision of

the concept of simultaneity into a more malleable form, I thus arrived at the special theory of relativity.' This simple summary may be comparatively easy to grasp (if we think it through), but the physical-mathematical argument and formulae involved in attempting to prove it are not. These Einstein now set down in a 31-page paper entitled 'On the Electrodynamics of Moving Bodies' (or in its historic German title: *Zur Electrodynamik bewegter Körper*).

To understand Einstein's Special Theory of Relativity (as he called it), one must first bear in mind the Newtonian system which it replaced. Indeed, for everyday purposes this Newtonian system is still very much how we continue to view the world. According to Newton everything from the orbiting planets to the falling apple are all subject to the same law: the force of gravity. The universe is seen as logical, its laws remaining consistent no matter where or under what circumstances they are applied. The foundations of this common sense world are space and time. As Newton put it so reassuringly in his *Principia*: 'Absolute, true and mathematical time, of itself and from its own nature, flows equably, without relation to anything external, and by another name is called duration.' Likewise 'absolute space, in its own nature, without relation to anything external remains always similar and immovable'. In other words, space and time are absolute. And so it seemed.

Whenever anyone had the temerity to question Newton on this matter, he referred them to God. This was just the way things were. It was simply how the universe had been ordained. But why? How did Newton know? It is the duty of scientific enquiry to ask such questions. But Newton's authority was so great that few dared. The attack was to come from a different front. Even when experimental evidence began to reveal discrepancies in the Newtonian account of the universe, initially few thought to question the entire edifice of classical physics.

Newton's classical physics dealt quite satisfactorily with relative motion. A sailor lying in his hammock might consider himself to be at rest with regard to his ship; but to someone on shore, viewing the ship under sail, the sailor would have a relative speed. Likewise, the stationary viewer on shore would acquire a great relative velocity if viewed from outer space, as he would take on the speed of the earth moving through space. But here the relativity stopped, for space was static and immovable (just like the elusive ether which filled it). This was the absolute standard of reference, along with absolute time.

Not until the 1860s was serious doubt cast on this state of affairs by Maxwell's electromagnetic wave theory of light (which had played such a central role in Einstein's paper on light). Maxwell's theory revealed a problem with Newton's classical mechanics when it came to the speed of light on moving objects. Could the velocity of light remain unaffected by the speed of the observer or the speed of its source? This seemed to be confirmed by the famous 1887 Michelson-Morley experiment to measure the speed of the earth through the ether. As we have seen, this cast doubt on the existence of the static all-pervasive ether. But it did much more than that. In essence, the intention was to measure the speed of light (s), and then measure the speed of light as it struck the earth in the direction of the earth's movement. This would be the speed of light minus the speed of the earth's movement (s-m). Subtract one from the other, and this would give the earth's speed: $s - (s-m) = m$. Yet astonishingly, the speed of light was found to be the same in both cases. The speed of the earth appeared to make no difference to the speed of light. But this couldn't be right. It defied common sense (to say nothing of Newtonian physics).

At around the same time Mach was beginning to question Newton's ideas of absolute space and absolute time. Mach's insistence on experimental evidence and facts reduced these to 'pure mental concepts that cannot be produced in experience'.

Before the turn of the century the greatest mathematician of the age, the Frenchman Poincaré, also cast doubt on the notions of absolute space and absolute time. Ingeniously, he argued that if one night whilst everyone was asleep the dimensions of the universe suddenly became a thousand times greater, the universe would remain utterly similar. How would we be able to tell what had happened? How could we measure this change in dimensions? We couldn't. The concept of space is thus relative to the frame of reference from which it is measured. Classical physics was approaching a crisis point, and Poincaré was well aware of this. He suggested: 'Perhaps we should construct a whole new mechanics where . . . the velocity of light would become an impassable limit.' Poincaré held back from such a step, which threatened to throw all scientific knowledge into chaos. But Einstein did not.

It was Einstein who finally discovered a solution to the many anomalies which had now been detected in classical physics. Einstein's achievement was to put forward a theory which not only accounted for these anomalies, but in the process proposed an entirely new explanation of the universe. In essence, he did this by proposing that the speed of light through space is constant – regardless of whether the source of light or the observer are moving. At the same time, he proposed that there is no such thing as absolute movement. This means there is no such thing as absolute rest. In which case, all speed is relative to its particular frame of reference (though the speed of light, being constant, will always be the same whatever the reference).

So far so good: the first proposal explained the Michelson–Morley experiment, and the second proposal explained away anomalies such as those pointed out by Poincaré. But as is quite evident, Einstein's two proposals seem contradictory. How can there be no such thing as absolute movement, if the speed of light is *always* the same?

Einstein now grasped the mettle. There was a way in which both these proposals could be true. This would mean accepting that both space and time were relative. But how could this be so? Poincaré had shown how space was relative; and hidden in his example of a thousand-fold increased universe was the implication that time too was relative. Einstein confirmed this, and confronted its awesome implications.

According to Einstein: 'all our judgements in which time plays a part are always judgements of *simultaneous events*. Take for example when I say: "That train arrives here at seven o'clock". What in fact I mean is something like: "The small hand of my watch pointing to seven and the arrival of the train are simultaneous events".' Einstein suggested that it might be possible to overcome these difficulties by simply substituting 'the position of the small hand of my watch' for the word 'time'. And this is adequate when we are talking only of the place where the watch is located. 'But it no longer holds good when we try to connect in time a number of events occurring at different places,' explained Einstein. 'Nor is it satisfactory for connecting the times of events occurring at places remote from the watch.'

Einstein always preferred theory to experiment. He also preferred reasoning to mathematics. The first quarter of his paper on the special theory of relativity is all but devoid of mathematical formulae, and these by no means make up the bulk of the latter parts. One of Einstein's great strengths lay in his ability to visualize in the simplest way complex mathematical situations. For instance, his thinking about relativity had been stimulated whilst he was travelling to work one day on the tram, gazing absent-mindedly back down the street at Berne's famous medieval Clock Tower. What would he have seen if the tram was travelling at the speed of light? According to the Special Theory of Relativity which he was developing, the clock on the tower would look as if it had stopped. Meanwhile the watch in his pocket would continue to move forward (though it would in fact move more slowly.

One consequence of Einstein's theory was that as velocity approached the speed of light, so time slowed down, becoming zero at the speed of light). Time just wasn't the same for every observer when speeds approached the speed of light.

Yet this prompts the obvious objection: but what about the 'real' time? The Clock Tower and the watch should obviously agree in 'real' time. But as Einstein had already argued, there is no such thing as 'real' time. Absolute time does not exist. Time only applies to the point at which it is being measured. *There is no other way it can be measured.*

This leads to some intriguing possibilities. Take the 'Twin Paradox'. One twin remains at home, while the other embarks on a long space journey at a speed approaching that of light. According to Einstein, when the astronaut twin returns he will be younger than his brother (his time will have slowed down throughout his journey, while the stationary twin will have continued at his 'normal' time).

After Einstein had finished his paper on the Special Theory of Relativity, he began working out its mathematical implications. These indicated some even more astonishing results, especially when the relativity principle was applied to the equations Maxwell had worked out for his electromagnetic theory of light. Einstein showed that when a particle travelled at a speed approaching that of light its mass increased, requiring ever vaster amounts of energy to propel it.

Around 1906 Einstein reached a crucial realization, which not only gave a further insight into the nature of quanta, but also pointed to an even more sensational development. It appeared that light quanta were simply particles which had somehow got rid of their mass and become a form of energy travelling at the speed of light. Mass, energy and the speed of light were somehow linked.

But now Einstein was to pay for the arrogance of his student years. He simply couldn't work his way through the mathematics involved. It took him almost two years − of great progress hampered by lack of technique and occasional blunders − before he finally came up with his celebrated formula encapsulating the link he knew existed. According to this: $e = mc^2$, where e is energy, m is mass, and c is the speed of light. This formula was literally earth-shattering. It implied that matter is solidified energy, and that if mass could somehow be converted into energy, a tiny amount of mass would release a stupendous amount of energy. The speed of light is approximately 300,000 kilometres per second. So if we convert Einstein's formula into $m = e/c^2$, this means that one unit of mass will release 90,000,000,000 units of energy.

This held the key to several questions which had troubled scientists for some time. For instance, it seemed to explain how the sun and the stars were capable of radiating such vast amounts of heat and light for millions of years. Somehow their matter was being converted into energy. But how? Experiments carried out by the Polish-French physicist Marie Curie in 1898 had discovered that one ounce of radium gave off 4,000 calories per hour indefinitely. Radium was a radioactive element; it was unstable and decayed to become radon, releasing energy in the process. Einstein's formula explained *what* happened; Madame Curie hinted *how* this happened. But it would be 25 years before Einstein's formula could even be verified. Einstein considered his famous formula to be the most important development resulting from his Special Theory of Relativity, but in those early days he had no idea how it could be put to use.

Back to 1905. Einstein finished his paper on the Special Theory of Relativity and posted it off to *Annalen der Physik*, where it duly appeared on 26th September 1905. Like any other young man who has just produced what he considers to be a work of consummate genius, he now sat back to await the world's amazed admiration. But such accolades are few and far between – as few and far between as genuine genius, though alas the two seldom coincide. And this was to be no exception.

For several months nothing happened. Had he made some simple miscalculation? But surely he couldn't have done this in all three of his major papers? Late summer became autumn, autumn became winter. Einstein once again began chopping up kindling wood and heaving sacks of coal upstairs for the smoking stove. Then in the New Year he received a letter from Max Planck asking for clarification of some of his calculations in the paper on relativity. Einstein knew at once that the significance of his work had been recognised by one of the great scientists of the time. Other recognition would surely follow. Yet despite this, it was slow in coming. Einstein's ideas were so revolutionary, and so contrary to common sense, that many wouldn't (or simply couldn't) take them seriously. It was no easy matter for physicists to accept the end of physics as they had known it.

Meanwhile Einstein continued to work at the Patent Office, occasionally stopping at a café to discuss his ideas with Besso over a coffee. The café happened to be a favoured haunt of the university scientific faculty, but Einstein remained unrecognised by the academics at the other tables. Einstein was now seeking to expand his theory of relativity so that it would account for gravitation. This was almost as complex and ambitious a task as the concept of relativity itself – but it concerned a topic on which he had thought long and hard for several years now.

One of those who were quick to recognise Einstein's work was Minkowski, his former mathematics professor at the Zurich Polytechnic (who had referred to him as a lazy dog). Indeed, Einstein's work was now beginning to suffer from his lack of mathematical application during his student years. The Special Theory of Relativity left many loose ends to be tied up and many avenues to be explored. Several of these were more mathematical than physical.

For a start, it became clear that three-dimensional geometry could no longer describe the universe. A new form of geometry was required. In 1907 Minkowski wrote a book called *Space and Time*, in which he made it clear that time should be treated as a fourth dimension. He demonstrated that neither time nor space could be viewed as having a separate existence. Time did not exist apart from the space to which it referred; likewise space did not exist except in time. The universe had to be seen as being constructed out of fused 'space-time'. Minkowski also produced the mathematics to back this up.

All this proved both an inspiration and a goad to Einstein. Others were now muscling in on his territory. But Minkowski's calculations gave him a profound insight. He suddenly saw how he could incorporate gravitation into relativity.

Newton had looked upon gravity as the force which attracted objects to one another. But what if objects moved instead in a gravitational field? Matter might then cause space to curve. Einstein regarded this inspiration as 'the happiest thought of my life'. The General Theory of Relativity was born – though it was to be over six years before work on this was complete.

At last Einstein managed to secure an academic post. But even now he needed the help of friends. His old student pal Adler, the political idealist, was appointed associate professor at the University of Zurich. But when Adler discovered that Einstein had also applied for the job, he selflessly resigned, commenting: 'If it is possible to obtain a man like Einstein for the university, it is absurd to appoint me.'

In 1909 the Einsteins moved back to Zurich, where their second son, Edouard, was born during the following year. Mileva felt more at ease back in the city where she had been a student, and Einstein took up his associate professorship. The students were at first perplexed by the sight of this scruffy young man, whose trousers were too short and whose hair was too long, standing diffidently beside the rostrum holding a crumpled visiting card. The visiting card turned out to be Einstein's lecture notes, but these he soon ignored – preferring to follow his train of thought. Afterwards he would invite his students to continue the discussion at the Café Terasse around the corner.

In 1911 Einstein was offered a full professorship at the German University in Prague. Here things remained much as before. When he first arrived at the gates the porter thought he was the electrician come to fix the lights. Einstein was pleased to be earning more money, but Mileva was deeply upset to leave Zurich. Mileva retired into her shell and Einstein affected not to notice, retreating into his work. Einstein's renown was now beginning to spread through the academic community and he was frequently away delivering speeches explaining his new theories. Mileva commented that he was away so much she was surprised he even recognised her.

On a lecture trip to Berlin in 1912, he met up again with his cousin Elsa Löwenthal, whom he had last seen 20 years previously in Munich. Elsa was 5 years older than Einstein: a comfortable 38-

year-old *hausfrau*, recently divorced, with two teenage daughters. She was motherly, rather than bright. Short-sighted, she read the newspapers pressed close to her face. Elsa was a practical woman with provincial attitudes who knew nothing about science. Not at all Einstein's type, it would seem – but she must have struck a chord. They began to correspond.

Possibly because of their childhood acquaintance, Einstein was unusually forthcoming right from the start. He told Elsa the usual old story about how his mother had never really loved him, and then added that he had always needed someone to love. Pretty soon he told Elsa that she was now fulfilling this role. Elsa seems to have responded in kind – but then Einstein began having misgivings. In her isolation Mileva had become a jealous woman. The letters between Albert and Elsa continued sporadically. She wrote to his place of work, and made him promise to destroy anything he received from her. Einstein may have played the absent-minded professor, but he didn't forget to burn Elsa's letters.

In 1914 Einstein was made Director of Physics at the Kaiser Wilhelm Institute in Berlin. He was

35 years old and academically he had at last arrived. The Kaiser Wilhelm Institute was one of the finest in the scientific world, and amongst several distinguished colleagues was Max Planck. Here Einstein could continue his researches undisturbed, and was only required to give the occasional lecture at the University of Berlin. To fulfil the requirements of the Institute he became a German citizen.

Mileva hated Germany even more than Prague. Within three months she had decamped to Zurich, taking the two children with her. The marriage appeared to have broken down irrevocably. Einstein was inconsolable at the loss of his sons. He sent the furniture from Berlin to furnish Mileva's Zurich flat, promised to send money every three months to support her, and settled down to a bachelor existence in his bare home. Elsa was living in the same district, and occasionally he would have a meal at her flat, but nothing more. He behaved as usual and buried himself in his work with a vengeance.

But this time there was something even Einstein couldn't ignore. In August 1914 the First World War broke out. Germany (like the other combatants throughout Europe) was swept with a frenzy of jingoism: columns of troops were cheered through the streets as they marched to the front – blissfully unsuspecting of the carnage that awaited them. Einstein was appalled. Even the Institute became involved. Some of Einstein's colleagues were commissioned to produce an efficient poison gas.

Einstein retired to his bare garret to continue working on his General Theory of Relativity, often not emerging for days on end. His few visitors spoke of uncarpeted floorboards. There were no books on the shelves; instead, scattered copies of the latest scientific journals and sundry sheets of paper covered with calculations littered the floor. Einstein himself often appeared at the door barefoot, and he appeared to sleep under an old rug. His hair was beginning to go grey, and it was now that it developed into the spectacularly unkempt mane so beloved of the cartoonists in his later years. A rare visitor described him as resembling: 'a bemused shaggy lion who has just suffered a huge electric shock'. Meals were infrequent, and simply prepared: everything cooked together in the same saucepan. When Elsa's daughter called in one day, she found him boil-ing an egg (its shell besmirched with chicken shit) in his soup. The effect on his digestive tract was predictable, and painful. On top of this was the emotional turmoil of his work, which brought him closer to breaking point.

This is hardly surprising. The work Einstein was engaged upon during this period has been described as: 'the greatest feat of human thinking about nature, the most amazing combination of philosophical penetration, intuitional physics, and mathematical skill'.

Einstein's previous *Special* Theory of Relativity had applied to bodies moving in relation to one another in uniform motion. The *General* Theory of Relativity extended to include bodies moving with accelerated relative motion, such as gravity (where a dropped object increases in speed). In order to do this, Einstein first had to scrap Newton's classical notion of gravity as a force act-ing between two bodies. Instead, he viewed grav-ity as an energy field which emanated from matter itself. The greater the amount of matter, the greater the effect of the gravitational energy it transmitted.

This may seem a minor matter, but the difference is crucial. Newton had based his entire universe upon a faulty conception of gravity. Newton's view of gravity as a force meant the sun's effect on the planets, and the planets' effects on their moons, were instantaneous. But as we have seen, according to Einstein's Special Theory of Relativity *nothing can travel faster than the speed of light.* Since the planets travel at around 1/1000 of the speed of light, the differences between calculations based on these conflicting views were infinitesimal. But differences they were: only one could be right. And the result was fundamental: only one could be the way the universe worked.

Einstein's view contained other, even more startling implications. Since 1905 Einstein had also extended his theory of light, developing the notion that it should be viewed as both particles and waves. But if light consisted of particles, it could be affected when it passed through a gravitational field. In other words, if light passed through a strong gravitational field it was liable to be curved.

But our entire notion of a straight line depended upon the passage of light. For instance, the shortest distance between two points in this curved field would not be a straight line. Like a plane flying the shortest distance between London and Los Angeles it would follow a curve.

Similarly our entire notion of ultimate speed (and thus space, and thus time) depended upon the speed of light. If a beam of light was bent when passing through a gravitational field, this meant that nothing could pass faster between two points on the curved beam other than *along* the curved beam. In other words there was *no* shorter distance between these two points, other than the curve. (This was what curved space *meant*.)

As a result, classical Euclidian geometry was no longer sufficient to describe the universe. Here was where Einstein's maths let him down. He could come up with no replacement. Without mathematical underpinning his theory was pure conjecture, and few conclusions could be drawn from it.

Fortunately for Einstein, work had been done on non-Euclidian geometry in the 19th century by the German Georg Riemann. For half a century his mathematics of curved surfaces had been considered utterly brilliant but utterly impractical. Riemann had shown that in curved geometry it was possible to draw any number of straight lines through two points. (Even a straight line from London through San Francisco will eventually pass through Los Angeles *on the globe*.)

Similarly Riemann demonstrated that in curved geometry there could be no such thing as a straight line of infinite length. Einstein realised that if space was curved, this also applied to the universe. A straight line would eventually meet up with itself again. Einstein's new conception of the universe was also greatly assisted by his old professor Minkowski's notion of space-time. This provided another link between the Special Theory and the General Theory, and tied up the loose ends left by the effect of curved light on space and time. Space became curved and so did time, which was not absolute but merely acted like a fourth dimension in a space-time continuum. (If light travelled in a curve, then time couldn't travel in a *faster* straight line, it had to travel in a curve too.)

Einstein published his results in March 1916 in *Annalen der Physik*, in an article entitled 'The Foundation of the General theory of Relativity'

(*Die Grundlage der allegmeined Relativitätstheorie*). Einstein's sensational new ideas were greeted with amazement and some bewilderment. It was all very well, but the whole thing was just theory. He claimed to be describing the universe, but all he provided was nothing but maths – with no practical proofs. Admittedly his theory seemed to explain a minute irregularity in the orbit of Mercury, for which Newton's physics could not account. But this was hardly conclusive practical proof for such enormous claims about the fundamental nature of the universe.

Einstein suggested a practical test. According to his theory, light from distant stars should be deflected when it passed through the strong gravitational field of the sun. Unfortunately such light could only be observed during an eclipse of the sun, and the next one wasn't due until 1919. The world would have to wait to discover whether it was part of a curved or a 'flat' universe.

Meanwhile the world found it had more important things to do. For most, March 1916 was marked by the massive slaughter of the Battle of Verdun. Einstein was horrified, and his pacifist views hardened.

When Einstein belatedly visited Switzerland, he realised that his marriage to Mileva was finished. On his return, he wept at the prospect of being separated from his two sons. Yet still he went ahead with divorce proceedings. The effect on Mileva was catastrophic, and she suffered a nervous breakdown.

All this pressure, coming after his long bout of extremely concentrated intellectual work, also brought Einstein to the point of collapse. According to his doctor: 'As his mind knows no limits, so his body follows no set rules; he sleeps until he is wakened; he stays awake until he is told to go to bed; he will go hungry until he is given something to eat; and then he eats until he is stopped'. Conditions in wartime Berlin were bad, and at one stage the disorganised Einstein lost four stone in two months. Elsa took him into her home to look after him.

The war finally ended in November 1918 with German defeat. The Kaiser fled to Holland, a socialist government took over, and political chaos ensued. Einstein was heartened by the socialist takeover and convinced that German militarism was now a thing of the past.

Under Elsa's motherly regime Einstein gradually recovered from his illness, but showed no sign of wishing to move back to his flat. He started working in his upper room, and in between the sound of him playing the violin would be heard. His meals were usually left outside his door. Einstein was oblivious to the domestic situation, and Elsa was only too pleased to let it continue. When his divorce came through, it was put to him that perhaps they should get married, and he seems to have gone along with this suggestion quite cheerfully. Elsa cut his hair, dressed him in a suit, and they were married in June 1919.

In November news came through that was to change Einstein's life for ever. Earlier in the year the British astrophysicist, Arthur Eddington, had led an expedition to the Portuguese African island of Principe in the Gulf of Guinea, where he had photographed the solar eclipse. Stars previously not visible because of the sun's radiance could now be observed. The photos also showed that as the stars' light passed close to the sun it was bent. That is, their position appeared to be different from when their light did not pass close to the sun. Eddington's observations matched Einstein's predictions about the light of distant stars being bent by the sun. The General Theory of Relativity was confirmed: for several days Einstein was in a state of euphoria.

But Einstein's reaction was nothing compared to that of the world's press. Very few really knew what relativity was about (even in scientific circles), but everyone understood that the universe had seemingly changed for ever. Suddenly the obscure physics professor from Berlin was being acclaimed as 'the greatest genius on earth'.

The world had just emerged from the catastrophic carnage of a world war, the so-called 'war to end all wars', and there was a widespread need for good news. The 'great men' of the past – military leaders, statesmen, aristocrats – had been discredited. The world was entering an age of populism ('the era of the common man') which needed to find its own new heroes. This process had begun in America with the phenomenon of Charlie Chaplin, and now Einstein was to join him. The unassuming absent-minded genius – who sometimes forgot to eat or put on his shirt, played the violin, and was liable to scribble formulas on the tablecloth when he came to dinner – was just what the press, and the public, were looking for. Likewise, post-war disillusion with the old forms of religion and philosophy had left a spiritual vacuum, which many tried to fill with relativity. Here was the *real* explanation of everything.

Einstein was now transformed into a public figure, travelling all over Europe giving public lectures explaining relativity. From Europe he travelled to America, where he was given the full ticker-tape treatment. ('Man who bent space visits Chicago.') Elsa would make sure he was properly dressed, and pretended to turn a blind eye to his flirtatious behaviour with society matrons. 'I am the person he goes home with,' she insisted, yet there were occasions when his behaviour caused her genuine distress. But this behaviour was more than simply a matter of personality: it was very much a matter of principle, too. Einstein's socialist principles included a belief in the utter freedom of the individual. His bohemian attitude extended to more than just his appearance.

In 1921 he won the Nobel Prize, and sent the $32,000 prize money to Mileva. He had secretly promised her this when they were divorced, some years before his work was recognised. Einstein had never doubted the importance of his achievements, or that one day his work would be recognised and rewarded.

Einstein was well aware of the ludicrous aspect of his celebrity. As a form of self-protection he played up his eccentricity (not a difficult task), but otherwise he insisted on putting his fame to good use. He lobbied long and hard for international disarmament, gave his full support to Zionism, and did all he could to counter the growing tide of anti-semitism in Germany.

In between his hectic lecture tours and campaigns, Einstein attempted to push his work forward. Although he had succeeded in redefining gravity and linking it to relativity, there were still loose ends to be tied up. Einstein wished to establish a mathematical relationship between electromagnetic forces (such as light) and gravity. This would lay the foundation for a fundamental law about the general behaviour of everything from the smallest electrons to the largest stars. Einstein was attempting to discover an even more fundamental formula than $e = mc^2$. He wished to relate all the properties of matter in a Unified Field

Theory. From this one absolute theory he would then go on to derive Quantum Theory. In this way he would be able to overcome the essentially ambiguous element of Quantum Theory, which defied logic by treating light as both waves and particles. As he famously insisted in a letter to the quantum theorist Max Born in 1926: 'I am convinced that God does not play dice'. But Nils Bohr, who was masterminding the development of Quantum Theory at Copenhagen, felt sure that Einstein's belief in a precisely engineered universe was mistaken. If anything, Quantum Theory was the absolute principle.

Einstein's published papers had always been greeted with scepticism – but from now on the scepticism would emanate from those who had previously supported him. He may have transformed the world, but it now looked as if he was being left behind. Einstein was an ambitious man, and this caused him some grief. Behind the facade of fame, these were hard times for Einstein. His son Edouard suffered a mental breakdown. Previously Edouard had hero-worshipped his father from afar, this now turned to hatred for deserting him and Mileva. Then the

Nazis took over in Germany, and offered a 20,000 mark reward for his assassination. 'I didn't know I was worth so much,' he commented, but was forced to flee to America.

Einstein accepted a permanent position at the Institute for Advanced Study in Princeton, which had recently been founded for pure research. When he arrived in America, he suddenly looked like an old man. Although he was only 54 his aureole of wild hair had gone completely white, and it was as if his face had turned to stone.

From now on, Einstein was to establish a routine which remained largely unchanged until the end of his life. Each morning he would leave his modest frame house at 112 Mercer Street and set off on the 20 minute walk to the Institute of Advanced Studies, which soon began to attract some of the finest minds in the scientific world.

This was the Einstein who became a living legend – the kindly eccentric-genius figure so beloved by the press. But in some ways he was now a sad figure. Einstein had long since parted company with his peers. Quantum Theory was now yielding spectacular results, and Einstein's insistence on searching for a Unified Field

Theory struck many as an utter waste of a supreme mind. As his friend Max Born remarked: 'Many of us regard this as a tragedy, both for him, as he gropes his way in loneliness, and for us, who miss our leader and standard-bearer'. Einstein had played a major role in the development of Quantum Theory, yet now he refused to believe its implications.

In 1936 Elsa died, and he withdrew further into his shell, working obsessively on his apparently futile calculations.

Einstein is said to have done two notable deeds during the last decades of his life. The first was sublime – both in its achievement and its horror. In 1939 the Danish physicist Nils Bohr called on Einstein at Princeton and confided some alarming news. Einstein's formula $e = mc^2$ had received dramatic confirmation. German scientists had split the atom and might soon be capable of constructing a bomb of unthinkable power. Einstein wrote informing President Roosevelt. Unbeknown to Einstein, Roosevelt secretly set up the Manhattan Project to manufacture the first atomic bomb. In 1945, when Einstein saw the results of what he had done, he instigated a world-wide campaign for the outlawing of nuclear weapons. For his troubles, he was investigated by the FBI.

By contrast, Einstein's second notable act of his last decades involved an element of the ridiculous. Einstein was now the world's most famous Jewish figure, and in 1952 he was offered the position of president of the newly-formed state of Israel. Never one to be taken in by his own myth, he declined with good grace.

Meanwhile Einstein had continued working on his Unified Field Theory. His herculean efforts persisted despite his failing health. One by one he was forced to give up his favourite pastimes. The chronic stomach pains (partly bequeathed by the singular dietary habits of earlier times) forced him to give up smoking his beloved pipe. Finally even his precious violin was set aside. But these were never major concerns. If anything, this abstinence now meant he had more time to concentrate on his one final aim.

In 1950 Einstein published a new version of his Unified Field Theory. It was greeted with an embarrassed silence by his scientific peers. He was now 71 years old, but had aged (in appearance at least) well beyond his years. He confessed that he often felt like a stranger in the world, but he felt at home enough to feel deeply disillusioned. The continuing FBI campaign against him, and his failure to resolve his Unified Field Theory were taking their toll. He became increasingly tired. In the spring of 1955, at the age of 76, he collapsed. Four days later, on 18th April 1955, he died in his sleep at Princeton Hospital. By his bed lay a page of unfinished calculations concerning his Unified Field Theory.

SOME KEY POINTS

Extract from Einstein's 1905 paper on relativity:

'The theory which will be developed is based – like all electrodynamics – on the kinematics of the rigid body. This is because the assertions of any such theory concern the relationships between rigid bodies (systems of co-ordinates), clocks and electromagnetic processes. Lack of attention to this fact is the cause of the difficulties at present faced by the electrodynamics of moving bodies.

1 Kinematical Part

Pt 1 Definition of Simultaneity

Take a system of co-ordinates where the equations of Newtonian mechanics hold good (ie, to the first approximation). In order to be precise and distinguish this system of co-ordinates from others to be used, we will call it the "stationary system".

If a material point is at rest relative to this system of co-ordinates, its position can be defined relative to the system by means of precise measurement and Euclidian geometry, and can be expressed in Cartesian co-ordinates.

If we wish to describe the *motion* of a material point, we give the values of its co-ordinates as functions of the time. However, we must realize that a mathematical description of this type has no physical meaning unless we are quite clear about what we understand by "time". We must realize that all our judgements in which time is included are always judgements of *simultaneous events*. For example, if I say: "That train arrives here at seven o'clock," what I mean is something like: "When the small hand of my watch points to seven, and the arrival of the train, are simultaneous events".'

Relativity in a nutshell:

Over vast distances time and space become relative. *Only the speed of light remains constant.*

Einstein's definition of relativity for the layman:

'When you are courting a nice girl an hour seems like a second. When you sit on a red hot cinder a second seems like an hour. That's relativity.'

The formula that led to the bomb:

$e = mc^2$

where e is energy released, m is mass, and c is the speed of light.

In the beginning:

As a result of Einstein's theory physicists have been able to trace the history of the universe to within a fraction of a second after the Big Bang. That fraction of a second has now been reduced to a decimal point, forty-two zeros, and a one – or .00000000000000000000000000000000000000000 00001. What happened at the precise instant of creation remains unknown, at least to science.

Some key sayings:

- 'I am convinced that God does not play dice.'
- 'The unleashed power of the atom has changed everything save our modes of thinking and we

thus drift toward unparalleled catastrophe.'

- 'I never think of the future, it comes soon enough.'
- 'Politics is for the present, but an equation is something for eternity.'
- 'If A is a success in life, then $A = x + y + z$. Work is x, y is play, and z is keeping your mouth shut.'
- 'If my theory of relativity is proven correct, Germany will claim me as a German and France will declare that I am a citizen of the world. Should my theory prove untrue, France will say that I am a German and Germany will declare that I am a Jew.'
- 'As far as the laws of mathematics refer to reality, they are not certain, and as far as they are certain, they do not refer to reality.'

Comments on Einstein:

- 'The genius of Einstein leads to Hiroshima.'

Pablo Picasso

- 'Einstein understands as much about psychology as I do about physics.'

Sigmund Freud

- 'Completely cuckoo.'

J. Robert Oppenheimer

CHRONOLOGY

Chronology of Einstein's Life

1879   Birth in Ulm, Germany

1894   Family move to Italy, leaving Albert in Munich, Germany

1895   Moves to Switzerland

1900   Graduates from Zurich Polytechnic Becomes Swiss citizen

1903   Marries Mileva Maric

1905   Publishes three ground-breaking papers, including one on Special Theory of Relativity

1909   Resigns from Patent Office in Berne

1913   Becomes Director of Physics at Kaiser Wilhelm Institute in Berlin

1916   Publishes paper on General Theory of Relativity

1919   Divorces Mileva and marries cousin Elsa Löwenthal

1919	Confirmation of Theory of Relativity brings world fame
1921	Awarded Nobel Prize for physics
1929	Publishes first version of Unified Field Theory
1933	Emigrates to United States after Nazi death threats
	Accepts full-time post at Institute for Advanced Study in Princeton
1939	Learns of splitting the atom, warns President Roosevelt
1940	Becomes a US citizen
1946	Branded 'commie stooge' over anti-nuclear stance
1950	Denounced by McCarthy
1955	Dies at Princeton aged 76

Chronology of the Age

1882   Death of Darwin

1889   Erection of Eiffel Tower in Paris

1900   Freud publishes *Interpretation of Dreams*

1903   Curies awarded Nobel prize for
       discovery of radioactivity

1907–14 Era of Cubism

1912   Sinking of the Titanic

1913   Stravinsky's *Rite of Spring* causes
       sensation in Paris

1914–18 First World War

1917   Bolshevik Revolution in Russia

1922   Publication of James Joyce's *Ulysses*

1929   Wall Street Crash heralds era of
       Depression

1933   Hitler comes to power in Germany

1939–45 Second World War

1945   Atomic bomb dropped on Hiroshima
       Founding of United Nations

1950   Outbreak of Korean War

SUGGESTIONS FOR FURTHER READING

Bernstein, Jeremy: *Einstein* (Modern Masters, 1996) – a good account of his ideas

Brian, Denis: *Einstein: a life* (Wiley, 1996) – the latest biography, giving unknown details re. Mileva

Clark, Ronald W: *Einstein: the life and times* (Hodder, 1988) – the standard full-length biography

Einstein, Albert: *Relativity: the special and general theory* (Methuen, 1935) – a popular exposition by Einstein himself

Einstein, Albert: *The Meaning of Relativity* (Methuen, 1932) – four lectures.

Michelmore, Peter: *Einstein: profile of the man* (Muller, 1962) – anecdotal picture of the man

Also in the Big Idea *series . . .*

PYTHAGORAS & HIS THEOREM

$$a^2 + b^2 = c^2$$

Most of us have heard about Pythagoras through his theorem on right-angled triangles, having been taught that the square on the hypotenuse is equal to the sum of the squares on the other two sides. But many are unaware of the significance of his theorem and the equation that neatly sums this up, and the enormous implications this has had for the way we view the world today.

Pythagoras was arguably the first mathematician and philosopher in the Western world. His work changed the way people viewed the world, establishing concepts such as abstract reasoning and deductive proof. *Pythagoras & His Theorem* presents a brilliant snapshot of Pythagoras' life and work, putting them in their historical and scientific context, and gives a clear and accessible explanation of their meaning and significance for the world we live in today.

The Big Idea is a fascinating series of popular science books aimed at scientists and non-specialists alike. Science is at its most exciting and gripping at moments of great discovery, and each of the books in the series looks in depth at the great moments that have advanced mankind's scientific knowledge and at the men and women who have made these huge breakthroughs in our thinking about the universe and our place in it.

CRICK, WATSON & DNA

Francis Crick's and James Watson's discovery of DNA – the very building blocks of life – has astounding implications for mankind's future. Not only in the scientific possibilities of cloning, life expectancy and medical research, but also in our everyday lives – in the genetic engineering of food and in forensics, for example. The discovery of DNA has also raised important ethical questions.

But what is DNA? What gateways has its discovery opened for future generations? And what of the sometimes frantic race that the scientists Crick and Watson were engaged in against other scientists to understand its construction and open up a whole new field of science? *Crick, Watson & DNA* presents a brilliant snapshot of these two scientists' lives and work, and gives a clear and accessible explanation of the meaning and importance of the discovery of DNA, and its implications for the twentieth century and beyond.

The Big Idea is a fascinating series of popular science books aimed at scientists and non-specialists alike. Science is at its most exciting and gripping at moments of great discovery, and each of the books in the series looks in depth at the great moments that have advanced mankind's scientific knowledge and at the men and women who have made these huge break-throughs in our thinking about the universe and our place in it.

TURING & THE COMPUTER

The computer has revolutionised the modern age of communication and information. It has touched every part of modern working life, to the extent that it would now be inconceivable without the computer. Its incredible power, and the speed it has brought to complex and multiple calculations, represent a massive leap forward in mankind's progress. Without doubt, the development of the computer is one of the twentieth century's greatest achievements.

But how many of us – even those of us who use a computer every day – know how it really works? And what of Alan Turing, the man who was a major figure in the development of the computer? The man who helped to break the Enigma codes during the Second World War, but who was largely forgotten after his death. *Turing & the Computer* presents a brilliant snapshot of Turing's life and work, and gives a clear and accessible explanation of the importance and meaning of the computer, and the way it has changed and shaped our lives in the twentieth century.

The Big Idea is a fascinating series of popular science books aimed at scientists and non-specialists alike. Science is at its most exciting and gripping at moments of great discovery, and each of the books in the series looks in depth at the great moments that have advanced mankind's scientific knowledge and at the men and women who have made these huge break-throughs in our thinking about the universe and our place in it.

☐ Pythago			£3.99
☐ Newton			£3.99
☐ Crick, Watson & DNA	Paul Strathern		£3.99
☐ Turing & the Computer	Paul Strathern		£3.99
☐ Hawking & Black Holes	Paul Strathern		£3.99

ALL ARROW BOOKS ARE AVAILABLE THROUGH MAIL ORDER OR FROM YOUR LOCAL BOOK-SHOP AND NEWSAGENT

PLEASE SEND CHEQUE/EUROCHEQUE/POSTAL ORDER (STERLING ONLY) ACCESS, VISA OR MASTERCARD, DINERS CARD, SWITCH, AMEX

☐☐☐☐☐☐☐☐☐☐☐☐☐☐☐☐

EXPIRY DATE .... SIGNATURE...................................

PLEASE ALLOW 75 PENCE PER BOOK FOR POST AND PACKING U.K.

OVERSEAS CUSTOMERS PLEASE ALLOW £1.00 PER COPY FOR POST AND PACKING.

ALL ORDERS TO:

ARROW BOOKS, BOOKS BY POST, TBS LIMITED, THE BOOK SERVICE, COLCHESTER ROAD, FRATING GREEN, COLCHESTER, ESSEX, CO7 7DW

NAME ....

ADDRESS..

....

Please allow 28 days for delivery. Please tick box if you do not wish to receive any additional information ☐

Prices and availability subject to change without notice.